CHEVROLET SPOTTER'S GUIDE
1920–1992

Tad Burness

First published in 1993 by Motorbooks International Publishers & Wholesalers, PO Box 2, 729 Prospect Avenue, Osceola, WI 54020 USA

© Tad Burness, 1993

All rights reserved. With the exception of quoting brief passages for the purposes of review no part of this publication may be reproduced without prior written permission from the Publisher

Motorbooks International is a certified trademark, registered with the United States Patent Office

The information in this book is true and complete to the best of our knowledge. All recommendations are made without any guarantee on the part of the author or Publisher, who also disclaim any liability incurred in connection with the use of this data or specific details

We recognize that some words, model names and designations, for example, mentioned herein are the property of the trademark holder. We use them for identification purposes only. This is not an official publication

Motorbooks International books are also available at discounts in bulk quantity for industrial or sales-promotional use. For details write to Special Sales Manager at the Publisher's address

Library of Congress Cataloging-in-Publication Data
Burness, Tad.
    Chevrolet spotter's guide 1920-1992 / Tad Burness.
      p. cm.
    Previous ed.: Chevrolet spotter's guide 1920-1980.
    Includes index.
    ISBN 0-87938-708-4
    1. Chevrolet automobile—Identification.   I. Title.
TL215.C5B86   1993
629.222'03—dc20   92-36784

All material in this book is reprinted from the following volumes:
American Car Spotter's Guide 1920-1939, American Car Spotter's Guide 1940-1965, American Car Spotter's Guide 1966-1980 and *American Truck Spotter's Guide 1920-1970.*

On the front cover: The 1949 Chevrolet woody wagon owned by Dennis Hollander. *David Gooley*

Printed and bound in the United States of America

Contents

Chevrolet Cars 1920–1929	4
Chevrolet Cars 1930–1939	9
Chevrolet Cars 1940–1949	14
Chevrolet Cars 1950–1959	19
Chevrolet Cars 1960–1969	30
Chevrolet Cars 1970–1979	60
Chevrolet Cars 1980–1989	105
Chevrolet Cars 1990–On	131
Chevrolet Trucks 1918–1949	136
Chevrolet Trucks 1950–1959	149
Chevrolet Trucks 1960–1969	152
Chevrolet Trucks 1970–1979	161
Chevrolet Trucks 1980–1989	172
Chevrolet Trucks 1990–On	191

for Economical Transportation
CHEVROLET

(FINAL 4-CYL. MODEL)

28

(AB) "NATIONAL"

Bigger and Better

HORSEPOWER INCREASED TO 35 @ 2200 RPM

WHEELBASE INCREASED TO 107"

NEW RADIATOR DESIGN AGAIN

for Economical Transportation
CHEVROLET

INTERIOR

4-WHEEL BRAKES

30 x 4.50 TIRES

The Chevrolet Special Sedan is a de luxe creation in every sense of the word. Standard equipment includes six wire wheels with fender wells, bumpers front and rear, robe rail, dome light, silk assist cords, etc.

30
"UNIVERSAL" (AD)

New Dash Gasoline Gauge

NEW DARK-FACED, CIRCULAR GAUGES

THE ROADSTER

THE SPORT ROADSTER

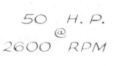

50 H.P. @ 2600 RPM

THE SPORT COUPE

THE COUPE

(CABRIOLET SUSPENDED FOR 1930; RE-INTRODUCED JANUARY, 1931.)

THE PHAETON

THE SEDAN

4.75 × 19" TIRES (THROUGH '31) AND NEW, SLIGHTLY SLANTED NON-GLARE WINDSHIELD →

THE CLUB SEDAN

(30½ MODEL HAS *Landau Irons*.)

THE COACH

9

CHEVROLET

$1741.

new "Bel-Air" 2-DR. HARDTOP has WIDE BACKLIGHT →

DASH

1950 TRUNK LID has new RE-DESIGNED HANDLE.

STYLING SIMILAR TO 1949, EXCEPT FOR MINOR DIFFERENCES AS NOTED.

new AUTOMATIC TRANSMISSION AVAILABLE ↓

HJ, HK

50

HJ = SPECIAL
HK = DE LUXE

new 1950 GRILLE WITHOUT VERTICAL LOWER CENTER PCS. SEEN IN '49.

First low-priced car with **POWERGlide** No-Shift driving *

PRICE RANGE: $1329. TO $1994.

* = POWERGLIDE SOMEWHAT LIKE BUICK'S "DYNAFLOW." (NOT INCLUDED IN ABOVE PRICES)

BACK SEAT (4-DR.)

The Styleline De Luxe 2-Door Sedan

1950 HUBCAP has YELLOW CENTER.

19

CHEVROLET

STYLELINE SPECIAL $1696.
(has MINIMUM of CHROME TRIM)

FLEETLINE DLX.
(NO MORE FLEETLN. SPECIAL)

NEW 52
KJ, KK

26 Exterior Colors and two-tone color combinations to choose from.

New Softer, Smoother Ride with new and improved shock absorber action.

STYLELINE DE LUXE 2-DR.

new 5 RIDGES RUN DOWN CENTER HORIZ. MEMBER of GRILLE.

Improved Carburetion with Automatic Choke in Powerglide models.

New Centerpoise Power is smoother — "screens out" engine vibration.

Color-Matched Two-Tone Interiors bring new beauty to De Luxe models.

new MEDALLION
$1519. TO $2281.
PRICE RANGE

STYLELINE DE LUXE SPORT COUPE (ABOVE)
(2 VIEWS)

$1992.

BEL AIR (in STYLELINE DLX. SERIES) H/T

FINAL YEAR FOR STYLELINE and FLEETLINE MODEL NAMES.

21

CHEVY II
100
$2313.
300
$2395.
$2710.
63
NOVA 400
new GRILLE
120 HP (6 CYL.) (SINCE '62)

NEW V8 POWER (OPTIONAL)
64 4, 6, OR V8
MORE '64s ON NEXT PAGE

(SINCE 1964) **CHEVELLE** BY CHEVROLET — MID-SIZE

FULL COIL SUSPENSION

115" WB

(300 IS LOWEST-PRICED, LITTLE CHROME, $2607 UP. 300 DLX. has CHROME STRIP ALONG SIDE.)

Malibu Wagon 2-SEAT

MALIBU CVT. = $3030.

MALIBU H/T SPT. CPE. $2821.

$3093.

194 CID 6 (120 HP)
230 CID 6 (140 HP)
283 CID TURBO-FIRE V8 (195 or 220 HP)
327 CID V8 (275 HP)
396 CID V8s (325 OR 360 HP)

FLUSH-AND-DRY ROCKER PANELS

66

(300 TYPES have PLAINER REAR DECKS without CHROME ORNAMENTATION)

DASH

OPTIONAL TACH.

SS 396 $3219.

SS has Turbo-Jet V8's (396 CID)

6.95/7.35 x 14 TIRES (OPT. RED-STRIPE NYLON TIRE and MAG.-STYLE WHEEL COVER)

(SS CVT. ALSO AVAIL.)

CHEVY II by CHEVROLET

(1962-1979)

DELUXE MODELS KNOWN AS **Nova**

153 CID 4 (93 HP)
194 CID 6 (120 HP)
283 CID V8 (195 HP)
6.50 x 13 TIRES
(WAG. = 6.95 x 14)

Station Wagon

110" WB (1962-1967)

NOVA H/ SUPER SPORT H/T $2652.

LOWEST-PRICED "100" 2-DR. = $2250.)

66 new GRILLE; new TALLER TAIL-LTS.

STATION WAGON AVAIL. THROUGH '67. (FROM $2709.)

NOVA SS H/T $2708.

4-CYL = 90 HP
6-CYL = 140 HP
V8 = 195 HP

67

1967 GRILLE DIFFERENCE ILLUSTR. AT LOWER LEFT

NOVA 4-DR. $2519.

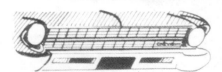

'67 Chevy II
The stylish economy car

Chevrolet CORVAIR

$2457.

a true hardtop. And it's Chevrolet's lowest priced hardtop.

Corvair 500

"500" INTERIOR

Corvair Power Teams

ENGINES	TRANS-MISSIONS	AXLE RATIOS
95 HP TURBO-AIR (164) Standard	3-Speed (Standard)	3.55:1
	3-Speed (Extra cost)	3.55:1
	Powerglide (Extra cost)	3.27:1
110 HP TURBO-AIR (164) Extra cost	3-Speed (Standard)	3.27:1
	4-Speed (Extra cost)	3.27:1
	Powerglide (Extra cost)	3.27:1
140 HP T. AIR (164) Extra cost	3-Speed (Standard)	3.55:1
	4-Speed (Extra cost)	3.55:1
	Powerglide (Extra cost)	3.55:1

$2840.
MONZA CONVERTIBLE

new DASH

1968 PRICES SHOWN. $10 INCREASE, 1969.
SALES:
1968 = 12,977
1969 = 3,102

'68 Standard Safety Features

new SIDE SAFETY LTS.

OPT. LUGGAGE CARRIER

(DISCONTINUED 5-14-69)

68-69

(FULL 1968-1969 LINE ILLUSTR.)

(NO 4-DR. HARDTOPS AFTER 1967)

MONZA INTERIOR

OPT. WIRE WHEEL COVERS

MONZA SPT. CPE.

$2721.

55

VEGA CHEVROLET

WAGON

GT.
2-DR. HATCH-BACK
(NEW)
BLACK GRILLE
A-70 x 13 TIRES
110 HP
(OTHER MODELS CONTINUE FROM '71)

'71-'72
INTRODUCING THE VEGA GT.

GT DASH (ABOVE) has ROUND GAUGES

CUSTOM INTERIOR

SWING-OUT REAR SIDE WINDOW (OPT.)

NO STYLING CHANGE, BUT SOME IMPROVEMENTS: EASIER-SHIFTING 3 and 4-SPEED TRANSMISSIONS. new WINDSHIELD WASHER CONTROL. NAMEPLATE NOW READS "VEGA BY CHEVROLET."

73

Better belts and jack.

STRONGER BUMPERS

new ESTATE WAGON $2850.

Chevrolet introduces a neat little woody.

CHEVY Nova

BACKGROUND SCENE: HISTORIC PLYMOUTH, MASS.

(RESTYLED) **73**

CPE. FR. $2589.
4-DR. FR. $2617.

250 CID 6 (100 HP)
307 CID V8 (115 HP)

Hatchback (New)
FROM $2738.

new 2-TONE ROOF ACCENT TRIM AVAILABLE (ON CAR ILLUSTR. ABOVE, CENTER)

4 TAIL-LIGHTS (new)

Chevrolet. Building a better way to see the U.S.A.

1973 DASH (BELOW)
(1974 DASH SIMILAR)

new GRILLE WITH PARK./DIRECTIONAL LIGHTS BUILT IN

NOVA/NOVA CUSTOM

MONZA

75½-76 (NEW) TOWNE COUPE

INTERIOR

"CABRIOLET ROOF"

new GRILLE and BODY

$3868.

Chevrolet makes sense for America.

Monza Towne Coupe
A small car and then some.

76

$3672. $3359 (SALE PR.)   (70 or 84 H.P. VERSIONS of 140 4 CYL. ENG.)

(Monza 2+2 $4040.

EPA 35 HIGHWAY 24 MPG CITY

5 YEAR 60,000 MILES
This 5-year 60,000-mile guarantee is an added value feature included in your 1976 Monza.

Get a little road magic.

$3875.

77

4 CYL. or 305 CID V8 AVAIL. (145 HP)

Monza Coupe.

SPYDER

Monza 2+2
2+2 for the road.

$4155.

Chevy Chevette

CHEVROLET — small car (1976 ~ 1987)

DELUXE DASH

1981 V.I.N. = 1G1AB089(-)B(-)000001 UP
SCOOTER = 7G1AJ089(-)B(-)100001 UP

COLUMN-MOUNTED "SMART SWITCH"

97.3" WHEELBASE ON THE 4-DR.

1981 MODELS ILLUSTR.

81-82

151 CID 4 or 173 CID V6
90 HP or 115 HP

98 CID 4
70 HP
12½ GAL. FUEL TK.

lower price!

$4595.00

REG. $5308.

"SCOOTER" 1TJ08
FROM $5730. ('82)

39 / 30 HWY. EST. / EPA EST. MPG

2-DR. (94.3" WHEELBASE)

P175/70 R13 TIRES (ROOF RACK OPTIONAL)

FILLER DOOR

New DIESEL MODELS ALSO AVAIL. '82. (116 CID 4 CYL.)

1982 V.I.N. = 1G1AB08C(-)C(-)000001 UP
(SCOOTER = 1G1AJ08C(-)C(-)000001 UP)

COMPUTER COMMAND CONTROL
Every 1981 gasoline-engined Chevrolet passenger car includes a sophisticated, thoroughly tested on-board computer as standard equipment.

AM RADIO (BELOW)

(AM/FM ALSO AVAIL.)

121

CHEVROLET CORVETTE.

BEST PRODUCTION SPORTS CAR IN THE WORLD.

(SINCE 1953)
(V8-POWERED SINCE 1955)
(ALL CORVETTES WITH FIBERGLASS BODIES)
98" WHEELBASE

P225/70R x 15 B TIRES

NEW AUXILIARY ELECTRIC FAN CUTS IN WHEN EXTRA ENGINE COOLING NEEDED. IMPROVED BATTERY.
24-GAL. FUEL TANK

1981
V.I.N. =
1G1AY8760
B(-)100001 UP

1YZ87

81

NEW FIBERGLASS-REINFORCED "MONOLEAF" REAR SPRINGS. NEW STAINLESS STEEL, FREE-FLOW EXHAUST MANIFOLDS.

4-SPEED MANUAL TRANSMISSION, OR AUTOMATIC.
350 CID V8 (190 HP)*
$16,141.

* CALIFORNIA 305 V8 ALSO (SINCE 1980)

New

STD. COUPE (1YZ87) $19368.
(P225/70R x 15/B TIRES)

V.I.N. =
1G1AY878(-)C(-)000001 UP

w. FRAMELESS GLASS HATCHBACK

COLLECT. ED. INTER AND EXTER. IN SILVER/BEIGE (METALLIC)

NEW "COLLECTOR EDITION" HATCHBACK COUPE (1YY07)

82-83

190 HP (OR 200)* WITH 350 CID V8

NO TRUE "1983" MODEL. CONTINUATION OF 1982 MODEL UNTIL 1984 READY (3-83)

$23,615.
P225/60R x 15/B TIRES

MALIBU CLASSIC

82-83

ALL 1982s ARE "CLASSIC."
1983s ARE SIMPLY "MALIBU."

NEW GRILLE WITH CRISS-CROSS PCS.

1982 V.I.N. ENDS WITH —C—#; 1983 " " —AXD—#

NO MORE 2 DR. MODELS

'83 HAS "MALIBU" NAME ON SIDE.

SPORT SEDAN $9321. ('82)   9335. ('83)

GAS or DIESEL V6s and V8s NOW AVAIL.

WAGON $9450.   9468. ('83)

("ESTATE WAGON" STILL AVAIL. ALSO)

1982 EXAMPLES ILLUSTRATED
21 GAL. FUEL TK. (18.2 GAL. WAGON)
21 MPG (EPA)
23 MPG DIESEL EPA

INTERIOR

1983 HAS "MALIBU" NAME ON SIDE, JUST BEYOND FR. WHEEL OPENING.

DISCONTINUED 1983

Malibu Classic instrument panel.

Malibu Classic standard notchback bench seat with folding center armrest.

CHEVROLET CAMARO

(SOME DATA OVERLAPS FROM PRECEDING CAMARO PAGE)

1986 (101" WB)
SPT. CPE. (V6) = $10914.
(FP 87 S)   173 CID
BERLINETTA SPT. CPE. (V6)
(FS 87 S) = $13531.
Z28 SPT. CPE. (V8) "
(FP 87 H)   305 CID

1987 (new CONVERTIBLES AVAIL.)
SPT. CPE. (FP 11 S) $11674.
RS CPE.   "
CONV'T. (FP 31 S) $16073. (V6)
"   (FP 31 H) $18897. (V8)
(V8 AVAIL. IN ALL BODY TYPES)

1988 ('86)
COUPE (FP 11 S) $12674.
RS   "
CVT. (FP 31 E) $17934.
IROC-Z SPT. CPE. (FP 11 E) $15169.
"   " CVT. (FP 31 E) $19694.

1989
RS CPE. (FP 11 S) $13199. (V6)
RS CVT. (FP 31 E) $18699. (V8)
IROC-Z SPT. CPE. (FP 11 E) $15849.
IROC-Z CVT. (FP 31 E) $20649.
(TPI 305 or TPI 350 V8s AVAIL.)

1990
RS CPE. (FP 23 T) $12754. (V6) (V8 OPT.)
RS CVT. (FP 33 E) $18639. (V8)
IROC-Z SPT. CPE.
(FP 23 F) $16374.
IROC-Z CVT. (FP 33 F)
$22014.
(WITH 5.7 L TPI V8, has
"8" Y.I.N. SUFFIX)

1991
RS CPE. (FP 24 T)
$13454. (V6)
RS CVT. (FP 33 T)
$19234. (V6)
Z28 CPE. (FP 23 F)
$17329. (V8)
Z28 CVT. (FP 33 F)
$22699. (V8)

Z28 CONV'T. ('91)

86-91

RS COUPE
(W. OPT. 16" ALUMINUM WHEELS)

RS CPE. $12075. (P215/65R15 TIRES ON RS)

(CAMAROS HAVE ALWAYS USED REAR WHEEL DR.)

Z28 CVT. $21500.
(P235/55R16 TIRES ON Z28)

(AVAIL. 350 CID V8)

WITH REAR SPOILER

T BAR ROOF OPT.

16" ALUMINUM WHEEL

92

Z28 COUPE $16055.

EPA MPG 16 CITY/25 HWY.
DRIVER'S SIDE AIRBAG

STD. DRIVER'S SIDE AIR BAG

RS CONVERTIBLE $18055.

15" RS Aluminum Wheel

DASH

CHEV. CAMARO 86~92

CHEVROLET

(TRUCKS SINCE 1918) BY CHEVROLET DIVISION OF GENERAL MOTORS CORP., DETROIT.

490	½ TON		
G	¾	120" WB	5.42 GR
T	1	125	

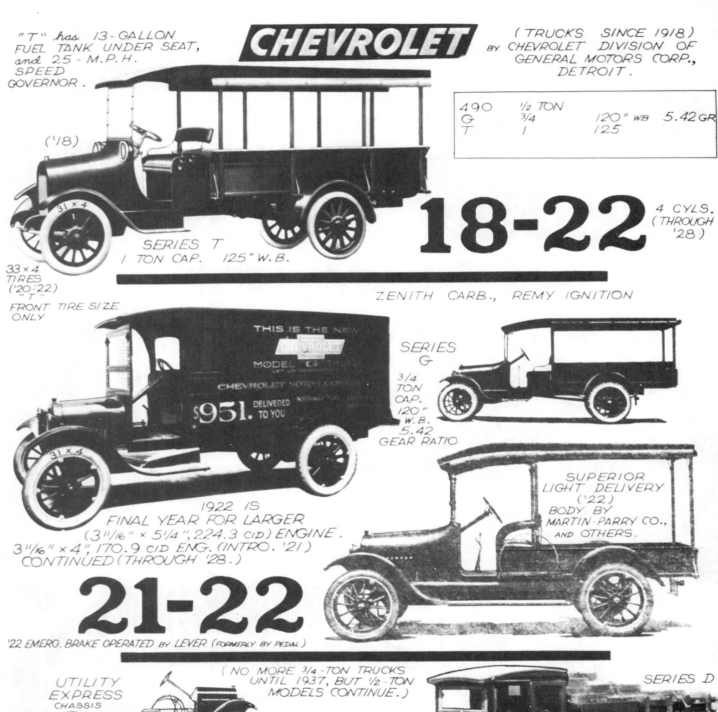

"T" has 13-GALLON FUEL TANK UNDER SEAT, and 25-M.P.H. SPEED GOVERNOR.

('18)

SERIES T 1 TON CAP. 125" W.B.

33 × 4 TIRES ('20-'22) "T" FRONT TIRE SIZE ONLY

18-22
4 CYLS. (THROUGH '28)

ZENITH CARB., REMY IGNITION

SERIES G ¾ TON CAP. 120" W.B. 5.42 GEAR RATIO

$951. DELIVERED TO YOU

1922 IS FINAL YEAR FOR LARGER (3 11/16" × 5 1/4", 224.3 CID) ENGINE. 3 11/16" × 4", 170.9 CID ENG. (INTRO. '21) CONTINUED (THROUGH '28.)

SUPERIOR LIGHT DELIVERY ('22) BODY BY MARTIN-PARRY CO., AND OTHERS.

21-22

'22 EMERG. BRAKE OPERATED BY LEVER (FORMERLY BY PEDAL)

UTILITY EXPRESS CHASSIS

(NO MORE ¾-TON TRUCKS UNTIL 1937, BUT ½-TON MODELS CONTINUE.)

SERIES D

1-TON

('23)

120" WB (THROUGH EARLY '25)

"D" "H"
23-24

136

for Economical Transportation
CHEVROLET

TRADITIONAL "BOW-TIE" EMBLEM

UTILITY EXPRESS ('24)

('23)

23-24 (CONT'D.)

"SUPERIOR" NAME CONT'D. FROM '24.
103" WB ON ½ TON CHASSIS
124" "   " 1-TON REPLACES 120" ('26.)

EARLY '25 IS MODEL "M."

25-26

LATE '25 "R" WITH HERCULES BODY

STAKE

('27)

LATE '26 MODEL "X" has NEW CHEVROLET-BUILT, FULLY ENCLOSED BODY.

"LM" HAS 1 TON CAP., 124" W.B., 30×5 TIRES ALL AROUND, AS DOES 1926 MODEL "X."

DELUXE 1-TON PANEL

OPEN EXPRESS

DASH

27 "CAPITOL"

new RADIATOR HAS DIP AT CENTER OF UPPER PAN.

137

CHEVROLET

ROOF-VISOR (LT. MODEL)

UTILITY TRUCK with 4-SPEED TRANSMISSION and 4-WH. BRAKES (8-28)

1927-1928 TRUCKS DO NOT HAVE STEERING COLUMN LOCK (AS USED ON '27-'28 CHEVROLET CARS.)

IN 1-TON, 124" W.B. MODELS, EARLY '28 IS "LO," LATER '28 IS "LP."

28

4-WHEEL BRAKES ON LT. DUTY MODELS

Illustrating the Light Delivery Chassis equipped with Panel Body

29 x 4.40 TIRES (30 x 5 REAR ON 1-TON)

CHEVROLET

('29)

SEDAN DELIVERY STYLING MORE LIKE THAT OF A CHEVROLET CAR.

NEW 6-CYL. ENGINE IN 1929

"INTERNATIONAL" "UNIVERSAL"

29-30

The Utility 1½ Ton Chassis with Chevrolet cab, equipped with power dump body built of reinforced steel to withstand concentrated weight. Popular among coal dealers, contractors, road builders, etc.

194 CID (THROUGH '32)

'29 : AC (½ TON) 107" WB (THROUGH '30)
LQ (1½ TON) 131" WB

STARTING 1930, SOME CANADIAN MODELS BEAR THE "MAPLE LEAF" NAME.

ALL MODELS HAVE 4-WH. BRAKES IN 1930.

1930 MODEL HAS new DASH with SMALL CIRCULAR GAUGES (INCLUDING ELECTRIC FUEL GA.) ILLUSTRATED AT RIGHT →

1½ Ton

POWER DUMP COMBINATION ('30)

The 1½ TON CHASSIS

50 HP and 4-SPEED TRANSMISSION

FINAL YEAR FOR THIS STYLE OF LIGHT DISC WHEELS.

STAKE ('30)

← SPARE TIRE and RIM MOUNTED HORIZONTALLY

DELIVERY MODELS OF 1930

'30 : AD (½ TON)
LR (1½ TON) (LATER BECOMES "LS") 131" WB
1½-TON CANOPY EXPRESS

SEDAN DELIVERY

PANEL DELIVERY

LIGHT DELIVERY

CHEVROLET

31 "INDEPENDENCE"

IND. COM. ½ TON 109" WB
Y UTILITY 1½ 131
UL DUAL 1½ 157

(ALSO KNOWN AS AE (½ T.)
LT, MA, MB, MC, MD
(1½ T.) MODELS)

50 HP @ 2600 RPM.
CARTER CARB.

4.75 × 19 TIRES
ON ½-TON.
(ALSO EARLY '32)

new LONGER GROUP OF MORE HOOD LOUVRES, SET IN SURROUNDING PANEL.

SEDAN DELIVERY (EARLY MODEL with TWO-BLADE BUMPERS)

LATER MODEL with SINGLE BLADE BUMPER

DISC WHEELS STILL AVAILABLE, BUT WIRE WHEELS TYPICAL AFTER 1930.

1½-TON OPEN EXPRESS (HEAVY PICKUP) with 157" WB. 30 × 5 TIRES

Open Cab Pick-up—Pick-up box 66 inches long, 45 inches wide and 13 inches deep. Body sides are so designed that they meet floor at right angles, permitting compact loading and generous capacity. Roadster-type cab. Disc wheels. Price of complete unit $440.

140

CHEVROLET

"EB" MODELS

- Sedan Delivery, $515 (107" Wheelbase)
- Half-Ton Pick-Up with Canopy, $495 (112" Wheelbase)
- Half-Ton Canopy Express, $555 (112" Wheelbase)
- Half-Ton Panel, $560 (112" Wheelbase)
- "QB" 1½-Ton Platform, $630 (131" Wheelbase)
- Half-Ton Pick-Up, $465 (112" Wheelbase)
- "QD" 1½-Ton Chassis and Cab, $605 (131" Wheelbase)
- 1½-Ton Chassis, $485 (131" Wheelbase)
- 1½-Ton Stake, $660 (131" Wheelbase)
- "QA" 1½-Ton Open Express, $655 (131" Wheelbase)
- "QD" 1½-Ton Stake, $720 (157" Wheelbase)
- 1½-Ton High Rack, $745 (157" Wheelbase)

IMPROVED BRAKING ON '35 MODELS
(1935 PRICES SHOWN)

NO MORE NAMES FOR YEAR MODELS.
34-35

'34 MODELS: DB, PA, PB, PC, PD

'35 MODELS: EB, QA, QB, QC, QD

COMMERCIAL	½ TON	112" WB	4.11 GR	(107" WB also)
UTILITY	1½	131	5.43	
"	1½	157	"	

CLOSER VIEW OF CANOPY EXPRESS ('34)

"BLUE FLAME" ENGINE
207 CID   70 HP @ 3200 RPM
CARTER CARB.

TIRE SIZES
5.50 × 17   (½ TON)
30 × 5 (FRONT) 32 × 6 (REAR)
(6.00 × 20)   (1½ TON)

"CARRYALL" METAL-BODY WAGON AVAIL. IN 1935.

CHEVROLET

36

- EARLY '36 CAB STYLE
- new INSTRUMENT PANEL SIMILAR TO THAT IN 1936 CHEV. "MASTER" SERIES CAR (THROUGH '39.)
- HYDRAULIC BRAKES ON ALL BUT EARLY 1½-TON "R".
- new HORIZONTAL HOOD LOUVRES
- SCREENSIDE CANOPY TRUCK

GEAR RATIOS
4.11 (½ TON)
5.43 – 6.17 (1½ TON)

- PANEL DELIVERY
- LATER '36 CAB STYLE 
- LATE '36 1½-TON PICKUP has STEEL ARTILLERY WHEELS.

½ TON: "FB" MODEL (112" WB)
1½ TON MODELS:
RA (131" WB)
RB (131" WB, DUAL REAR WHEELS)
RC (157" WB)
RD (157" WB, DUAL REAR WHEELS)

- HEAVY-DUTY
- FB LIGHT-DUTY ½-TON PICKUP 
- SEDAN DELIVERY HAS GRILLE LIKE 1936 CHEVROLET CAR.

CHEVROLET

IMPROVED DIAPHRAGM-SPRING CLUTCH, VOLTAGE-REGULATOR GENERATOR

½ TON PICKUP

C.O.E. (RARE)

CAB ('38)

HC, HD, HE, TA, TB, TC, TD MODELS

38

1½-TON STAKE

new TRUCK GRILLE with HEAVIER HORIZONTAL MEMBERS

39

JC (½ TON;)
JD (¾ TON;)
JE (1 TON;)
VA, VB, VC, VD (1½ TON;)
VE, VF, VG, VH, VM,
VN (1½ TON C.O.E.)

78 HP @ 3200 RPM
113½" WB and up

45 MODELS, 8 WHEELBASES IN 1939.

new V-WINDSHIELD ON ALL 1939 MODELS.

New Chevrolet-Built CAB-OVER-ENGINE MODELS

UNLIKE PANEL TRUCK, SEDAN DELIVERY IS STYLED LIKE CHEVROLET CAR.

TOP PIECE OF GRILLE IS VERT. WIDENED

new HYPOID REAR AXLE
FULL-FLOATING REAR AXLE ON HEAVY-DUTY MODELS with VACUUM-POWER BRAKES and 2-SPEED REAR AXLE OPTIONS.

40

CHEVROLET

new SEALED-BEAM HEADLIGHTS, and PARKING LIGHTS ON FENDERS.

new 4.55 GR ON ¾ TON and 1-TON

C.O.E. (LOGGER)

MODELS: KC (½ TON;) KD, KE (¾ TON;) KF (1 TON)
KP (½-TON PARCEL DELIV.;)*
WA, WB (1½ TON;)
WD, WE, WF (1½ TON C.O.E.)

* = SPRING INTRODUCTION

(1940 CHEVROLET CAR-TYPE INSTRUMENT PANEL REPLACES 1936 TYPE (EXCEPT IN SCHOOLBUS AND FLAT-FACE COWL TYPE.)

BEFORE WORLD WAR II, ALL GM-BUILT TRUCKS OVER 1½-TON CAP'Y. SOLD UNDER GMC NAME.

"WC" IS 1½-TON SCHOOL BUS CHASSIS.

HEAVY-DUTY DUMP TRUCK

CHEVROLET 41-47

new 90-H.P. and 93-H.P. ENGINES ('41)

60 MODELS, 9 WHEELBASES ('41) (SOME MODELS ARE LISTED BELOW.)

WARTIME MODELS HAVE LESS CHROME and NO FLOOR MAT.

216 OR new 235 CID, STARTING 1941.

PICKUP

MILITARY TRUCKS ('43)

1941 MODELS: AK, AJ (115" WB;) AL (125" WB;) AN, YR (134" WB;) YS (160" WB) C.O.E.s = YU, YV, YW

1942 MODELS: BK, BJ (115" WB;) BL (125" WB;) (CONT'D. TO '45) BN, MR (134" WB;) MS (160" WB) C.O.E.s = MU, MV

1946 MODELS: CK (115" WB;) OR, OE (134" WB;) (INTRO. '45) OS, OF (160" WB) C.O.E.s = OH, OI, OJ

1947 MODELS: DP (115" WB;) DR (125" WB;) DS, PJ, PV(S,) (134" WB;) PK, PW(S)(160" WB) C.O.E.s = PP, PR, PS TANKER (C.O.E.)

2-TON MODELS AVAIL. IN 1946.

VARIETY OF MODELS ILLUSTRATED ('45)

CHEVROLET
Advance-Design
(TOTALLY RESTYLED)
INTRO. SUMMER, 1947

STAKE ('48)
PICKUP ('51)

90 OR 93 HP IN 1948
HP INCREASED TO 92, 102, 105 IN 1950.
92, 107, 108 HP IN 1953.

48-53

YEAR DETERMINED BY FIRST LETTER IN MODEL DESIGNATION:

E, Q = 1948
G, S = 1949
H, T = 1950
J, U = 1951
K, V = 1952

WIDE VARIETY OF ALPHABETICAL SINGLE-LETTER MODEL DESIGNATIONS IN 1953.

HEAVY PANEL ('51)

LIGHT PANEL TRUCKS (SEDAN DELIVERY) STYLED LIKE EACH YEAR'S NEW CHEVROLET CARS.

C.O.E. ('49)

"ADVANCE-DESIGN" NAME CONT'D.

CHEVROLET

DRIVER'S COMPARTMENT OF PANEL TRUCK

THREE 6-CYL. ENGINES: "THRIFTMASTER 235," "LOADMASTER 235," OR "JOBMASTER 261"

'54

FIRST RESTYLING SINCE '48.

CAB INTERIOR OF CAB-OVER-ENGINE

NEW 1-PC. CURVED WINDSH.
NEW GRILLE

55-56

NEW 12-VOLT IGNITION

CHOICE OF 6-CYL. OR NEW V8 ENGINES

NEW "CAMEO" SPT. PICKUP has BROAD FIBERGLASS BOX (THROUGH '57.)

TUBELESS TIRES ON 1/2-TON

CAB-FORWARD

('56)

new INDEPENDENT FRONT SUSPENSION *with* TORSION BARS *and* BALL JOINTS IN FRONT, COIL SPRINGS AT REAR (ON TRUCKS TO 3/4 TON CAP.)

(CONT'D.) **60**

INSTRUMENTS (IN TRUCKS UP TO 2-TON CAP'Y.)

new "TILT-CAB" MODEL

STURDI-BILT TRUCKS CHEVROLET

Rotary Valve Power Steering
AVAIL. ON 60, 70 and 80 SERIES

154

PROD.: 394,017

1961 ENGINE SPECS.

TURBO-AIR	6 CYL.	145 CID	80 HP @ 4400 RPM
THRIFTMASTER	6	236	135 @ 4000
JOBMASTER (SP.)	6	261	150 @ 4000
WORKMASTER (SP.)	V8	348	185 230 HP

FRONT END DETAILS

ENGINES 6 V8

61

PICKUP CAB

(WIDTHS EXAGGERATED)

60 SERIES (Middleweight)

70 SERIES

PROD.: 342,659

CORVAIR 61-64

(TRUCKS — 1961–1964) BY CHEVROLET DIVISION OF G.M.

AIR-COOLED REAR-ENGINE 6 CYLS. 80 H.P. 8 TO 1 COMPR.

"CORVAN" PANEL TRUCK

"GREENBRIER" SPORTS WAGON

CORVAIR 95'S (95" WHEELBASE)

"RAMPSIDE" PICKUP ('61) ("LOADSIDE" MODEL DOES NOT HAVE SIDE RAMP.)

(CORVAIR CARS PRODUCED 1960–1969.)

CAB ('61)

('64)

1ST DIGIT IN SERIAL NUMBER DETERMINES YEAR:
"1-R124" ('61)
"2-R124" ('62)
"3-R124" ('63)
ETC.

ADDITIONAL VIEWS OF UNIQUE "RAMPSIDE"

CHOICE OF 18 LIGHT-DUTY, 258 MEDIUM and HEAVY-DUTY MODELS IN 1965!

HEAVY-DUTY STYLING UNCHANGED

STAKE

CHEVROLET

PROD.: 619,690

65

4, 6, and V8 GAS ENGINES, (also FOUR DIESELS.)

VARIOUS OTHER 1965 TYPES ILLUSTRATED BELOW:

SEEN FROM FRONT TO REAR: CHEVY-VAN, STEP-VAN 7, 60 SERIES TRUCK with VAN BODY, 80 SERIES DIESEL TRACTOR

(½-TON PANEL ACROSS THE STREET.)

Chevrolet

PRODUCTION: 492,601

CHEVY VANS START AT VEHICLE I.D. # GS-150 (-) 100001 FOR 1970.

$2489. UP

Chevy Van

STD. FLEETSIDE PICKUP (ABOVE,) SEEN WITH OPTIONAL CAMPER.

PICKUPS FROM 115" WB 250 CID 6 (155 HP) OR 307 CID (200 HP) V8, PLUS 3 OTHER ENGINE OPTIONS.

Pickups:
FROM $2654.

70 new GRILLE (12 SETS OF SMALL HORIZONTAL FINS)

new VAN DESIGNS (DETAILS AT UPPER RT.)

250 OR 292 CID 6, 307, 350, OR new 400 CID V8 AVAILABLE IN PICKUPS.

3 OR 4-SP. MANUAL TRANS. OR 2-SP. POWERGLIDE OR 3-SP. TURBO Hydra-Matic AVAIL.

CHEVROLET On the move.

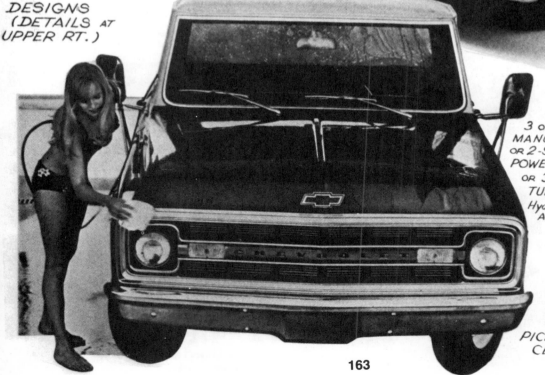

A cargo box inside a cargo box.
One reason behind Chevy's longer life is an extra wall of steel behind every Fleetside box wall. Inside dents stay inside. Sleek outside styling stays sleek.

PICKUP I.D.# STARTS AT CE 140 (-) 100001

Chevrolet

CHEVY-VAN I.D. NOS.:
'73 = CG(Q)153(F)*
100001 UP
*= 154(F) '74

Chevy Van.
FROM $2818.
(3092. '74) (SEE ALSO GMC **VANDURA**)

73-74 (CONT'D.)

Aluminum rust-free body.

Step-Van
(note DIFFERENCES BETWEEN ←ALUMINUM and STEEL→ BODY FLOOR, HOOD, INNER WALLS, ETC.

All-steel Step-Van King bodies.

1974 PARCEL DLVY. VAN I.D. NOS. START AT CC(Q)154(F) 100001

Special window combinations are available. These include windows in the rear doors, sliding door, rear side panels or any combination of the three.

1973 EXAMPLES ILLUSTR.

Economical 250 Six standard.

Building a better way to serve the U.S.A.

CHOICE of STEEL OR ALUMINUM BODIES IN STEP-VAN 7 ALSO, STARTING 1974.

Built like the big Step-Vans.

Special van bodies
Special van bodies that fit the Chevrolet Forward Control Chassis are designed and built by several body manufacturers. These special bodies give you an even wider van selection with all the benefits of Chevrolet chassis quality. Major special body manufacturers include the following companies: J.B.E. Olson Corp., Garden City, N.Y.; Boyertown Auto Body Works, Boyertown, Pa.; Union City Body Co., Inc., Union City, Ind.; Lyncoach and Truck Co., Oneonta, N.Y.; Penn Versatile Van, Chicago, Ill.; Fixible Southern Co., Evergreen, Ala.

IN 7', 8' OR 1974 ALUM. 9'

Step-Van 7 with new extended hood design for quick easy service.

Chevy LUV

Chevrolet's imported Light Utility Vehicle

Chevrolet LUV produced by Isuzu Motors Limited, Japan.

IMPORTED (SINCE MARCH, 1972)
4 CYL. 110.8 CID
75 HP (THROUGH '79)
FROM $2196.   102.4" WB

INTERIOR

NEW

72

6.00 x 14/C TIRES (THROUGH '76)

1500 lbs. payload capacity.
102.4" wheelbase.
Power brakes standard.

$2406. ('73) SERIES 2 $2707. ('74) SERIES 3

73-74

little Chevy import big Chevy value.

Stowage area behind seats.

START WITH '74 MODEL, new LUXURY "MIKADO" INTERIOR OPTION.

Crank-down spare tire.

new TAIL-LIGHTS (EARLY '73)

LUV's spare tire is tucked up beneath the cargo box. A convenient crank inserted on the passenger side raises and lowers the spare from the stowage position by chain and winch.

CHEVY LUV

WITH 4-W-D STRIPING
FR. $4401. ('79)
('79)
23 EPA ESTIMATED MPG
32 ESTIMATED HIGHWAY

CHEVY LUV — BUILT TO STAY TOUGH

('80) new 80 HP ('80)

ANNOUNCING 4-WHEEL-DRIVE (OPTIONAL)

79-80

LUV SERIES 10 **22** EPA EST MPG **30** EST HWY ('80)
4WD SERIES 9   SERIES 10 FR. $4935.

AN IMPORTANT STEP AHEAD

LUV SERIES 11 SPECIFICATIONS		
MODELS	(2WD)	(4WD)
WHEELBASE (in.)	104.3 / 117.9	104.3
GVWR (lbs.)	3550 / 4150	3750
CURB WEIGHT (lbs.)	2375 / 2470	2595
GROSS PAYLOAD (lbs.)*	1175 / 1680	1155
ENGINE—Type (code)	OHC-4 cyl (L10)†	
Displacement	1.8 Liter (110.8 Cu. in.)	
Net H.P. (SAE)	80 @ 4800	
Net torque (SAE)	95 @ 3000	
Comp. ratio	8.5 to 1	

new DASH ('81)

Mikado instrument panel. Includes special bright accents around the instrument panel, cluster and vent grilles of optional Mikado decor.

('81)

E78 × 14 TIRES STD. (SINCE '77) F70 × 14 (4WD)

New wheel cover. Optional.

81-82

RESTYLED
new 104.3" WB
NEW GRILLE
3 REAR QUARTER LOUVRES REPLACED BY new MEDALLION.
new 4-CYL. 136.6 CID DIESEL ENG. OPT. IN 1982

Sliding rear window OPT.

Fiberglass cap. Helps protect cargo from the weather. Dealer installed.

FR. $6039. ('81)
SERIES 11 ('81)
SERIES 12 ('82)

If you have enjoyed this book, other titles from Motorbooks International include:

75 Years of Chevrolet, by George H. Dammann
Chevrolet Small-Block V-8: History of the Chevrolet Small-Block V-8 and Small-Block Powered Cars, by Anthony Young in the Motorbooks International Muscle Car Color History Series
Chevrolet SS Muscle Car Red Book, by Peter C. Sessler in the Motorbooks International Red Book Series
Tri-Chevy Red Book, by Peter C. Sessler in the Motorbooks International Red Book Series
Chevrolet Super Sports 1961-1976, by Terry V. Boyce
Chevrolet Chevelle SS Restoration Guide, by Paul Herd in the Motorbooks International Authentic Restoration Guide Series
Chevrolet Chevelle & Monte Carlo 1964-1972, by James H. Moloney in the Classic Motorbooks Photofacts Series
Chevrolet El Camino 1959-1982, by Donald F. Wood in the Classic Motorbooks Photofacts Series
Illustrated Chevrolet Buyer's Guide, by John Gunnell
Illustrated Corvette Buyer's Guide Third Edition, by Michael Antonick
Illustrated Camaro Buyer's Guide Second Edition, by Michael Antonick
Illustrated Chevrolet Pickup Buyer's Guide, by Tom Brownell
Chevrolet '55-'56 Restoration Guide, by Nelson Aregood, Wayne Oakley & Joe Umphenour
Chevrolet '57 Restoration Guide, by Nelson Aregood, Wayne Oakley & Joe Umphenour
Chevrolet SS Restoration Guide, by Paul A. Herd
Pickup & Van Spotter's Guide 1945-1992, by Tad Burness

Motorbooks International titles are available
through quality bookstores everywhere.

For a free catalog, write or call
Motorbooks International
P.O. Box 1
Osceola, WI 54020
1-800-826-6600